I0481404

Contents

CHAPTER-1

What is FTTx Network?

FTTx, also called as fiber to the x, is a collective term for any broadband network architecture using optical fiber to provide all or part of the local loop used for last mile telecommunications. With different network destinations, FTTx can be categorized into several terminologies, such as FTTH, FTTN, FTTC, FTTB, etc.

FTTx provides an excellent platform for high or ultra-high speed access technologies. Not only do fixed access networks benefit from FTTx solutions, but advanced wireless networks do as well, especially in regard to increased backhaul capacity. While the move from copper-based networks to FTTx implies a big change for operators, the challenges involved in the deployment and operation of FTTx have been addressed with a multitude of proven solutions for both the passive as well as the active parts of the network. FTTx is now a reality with more than 100 million subscribers the world over

FTTH

FTTx is commonly associated with residential FTTH (fiber to the home) services, and FTTH is certainly one of the fastest growing applications worldwide. In an FTTH deployment, optical cabling terminates at the boundary of the living space so as to reach the individual home and business office where families and officers can both utilize the network in an easier way.

FTTN
In a FTTN (fiber to the node) deployment, the optical fiber terminates in a cabinet which may be as much as a few miles from the customer premises. And the final connection from street cabinet to customer premises usually uses copper. FTTN is often an interim step toward full FTTH and is typically used to deliver advanced triple-play telecommunications services.

FTTC
In a FTTC (fiber to the curb) deployment, optical cabling usually terminates within 300 yards of the customer premises. Fiber cables are installed or utilized along the roadside from the central office to home or office. Using the FTTC technique, the last connection between the curb and home or office can use the coaxial cable. It replaces the old telephone

service and enables the different communication services through a single line.

FTTB

In a FTTB (fiber to the building) deployment, optical cabling terminates at the buildings. Unlike FTTH which runs the fiber inside the subscriber's apartment unit, FTTB only reaches the apartment building's electrical room. The signal is conveyed to the final distance using any non-optical means, including twisted pair, coaxial cable, wireless, or power line communication. FTTB applies the dedicated access, thus the client can conveniently enjoy the 24-hour high speed Internet by installing a network card on the computer.

♣ MDU – Multiple dwelling units.

♣ MTU – Multiple tenant units.

Benefit of FTTx

FTTx improves efficiency by delivering cable services over fiber optics.

- Fiber to the premises is extremely cost effective in rural areas
- Improved overall reliability and reduced signal egress

- No power required for active devices and battery power supplies in the outside plant (OSP)
- Backup power for consumer premises equipment becomes the responsibility of the subscriber

Applications on FTTx:

- IP video services like IPTV
- Over the top (OTT) services like video consumption
- Increased use of HD video
- UHD 4K video w/ HDR
- Internet of Things (IoT)
- Security
- VOIP
- RF video

Questions

1. What is FTTH?
2. What is FTTB?
3. What are benefits of FTTx?
4. Write any five applications on FTTx?

CHAPTER-2

FTTH Network Architecture

2.1 The FTTH network environment

The deployment of fibre closer to the subscriber may require the fibre infrastructure to be located on public and/or private land and within public and/or private properties.

- Population density : Each physical environment constitute different subscriber dwelling densities (per sq km)

  I. City Area
  II. Multi dwelling Unit
  III. Open Residential or single dwelling unit
  IV. Rural Area

- Type of FTTH Site : The nature of the site will be a key factor in deciding the most appropriate network design and architecture

I. Brown Field: buildings are already in place but the existing infrastructure is of a low standard
II. Green Field: New build where the network will be installed at the same time as the buildings
III. Over build: adding to the existing infrastructure

- Deployment Technology: The fiber deployment method and technology will determine CAPEX and OPEX, as well as the reliability of the network. These costs can be optimized by choosing the most appropriate active solution combined with the most appropriate infrastructure deployment methodology.

I. Conventional Duct
II. Micro Duct
III. Aerial
IV. Direct Buried
V. RoW (Right of Way)

Drop point can be set at free positions without creating any fixed loops beforehand. You can install closures anytime anywhere you want.

2.2 FTTx Technology

Passive optical network (PON) based FTTx access network is a point-to-multipoint, fiber to the premises network architecture in which unpowered optical splitters are used to enable a single optical fiber to serve multiple premises.

2.3. FTTx Access Network Architecture

GPON's have a tree topology in order to maximize their coverage with minimum network splits, thus reducing optical power. An FTTH access network comprises five areas, namely a core network area, a central office area, a feeder area, a distribution area and a user area. The core network area is not considered as a part of the FTTH access network. The network architecture represents two level of splitting between the central office and the user premises achieving an overall splitting ratio of 1:64. The distance between the OLT and ONT may be more than 20 km depending on the total available optical power budget, which is a factor of the OLT laser port and the total loss budget.

9

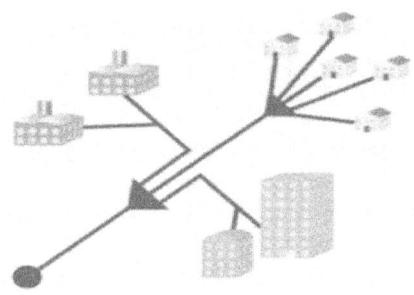

Point to Multi-Point

2.3.1 **FTTH Core Network**

The core network includes the internet service provider ISP equipment's e.g. DWDM, Routers etc

2.3.2 **Central Office:** The main function of the central office is to host the OLT and ODF and provide the necessary powering. Sometimes it might even include some (or all) of the components of the core network.

Parameter	BPON	EPON	GPON	XGPON	10G-EPON
Standard	ITU-T G.983	IEEE 802.3ah	ITU-T G.984	ITU-T G.987	IEEE 802.3av
Downstream data Rate	622 Mbps	1.25 Gbps	2.5 Gbps	10 Gbps	10 Gbps
Upstream data Rate	155 Mbps	1.25 Gbps	1.25 Gbps	2.5 Gbps	10 Gbps/Symmetric 1 Gbps/Asymmetric

TDM PON standards

2.3.3 **FTTH Feeder Network**

The feeder area extends from optical distribution frames (ODF) in the central office CO to the distribution points. These points are usually street cabinets, where level-1 splitters are place. . The feeder cable is usually connected as ring topology.

2.3.4 FTTH Distribution

Network Distribution cable connects level-1 splitter with level-2 splitter. Level-2 splitter is usually hosted in a pole mounted box called Fiber Access Terminal FAT usually placed at building floor or the entrance of the neighborhood.

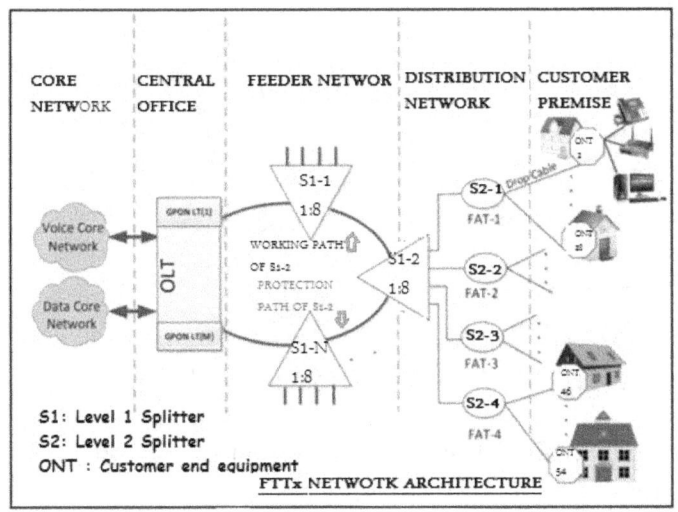

FTTx NETWOTK ARCHITECTURE

11

2.3.5 Drop Cable

In the drop cables are used to connect the level-2 splitter inside the FAT to the subscriber premises. Drop cables have less fiber count and length ranges up to 100 meters. Drop cables are designed with attributes such as flexibility, less weight, smaller diameter, ease of fiber access and termination. For ease of maintenance, usually an aerial drop cable is used.

Component of GPON Technology:

2.4.1 Optical Line Terminal OLT The Optical Line Terminal (OLT) is the main element of the network. OLTs typically operate using redundant DC power (-48VDC) and have at least 1 Line card for incoming internet, 1 System Card for onboard configuration, and 1 to many GPON cards. Each GPON card consists of a number of GPON ports.

Function of OLT :
- OLT perform are traffic scheduling,
- buffer control and bandwidth allocation

2.4.2 **Optical Splitters** The optical splitter splits the power of the signal that is each link (fiber) entering the splitter may be split into a given number of fibers leaving the splitter and there is usually three or more levels of fibers corresponding to two or more levels of splitters. This enables sharing of each fiber by many users. Due to power splitting the signal gets attenuated but its structure and properties remain the same. Splitters are a main component of GPON optical distribution networks

The passive optical splitter need to have the following characteristics:

- Broad operating wavelength range
- Low insertion loss and uniformity in any conditions
- Minimal dimensions
- High reliability

Types of Splitter

- Fused Biconical Taper (FBT)

I.   FBT splitters are made by fusing together two wrapped fibres.
II.   Monolithic devices are available up to 1x4 split ratio.
III. Split ratios greater than 1x4 are built by cascading 1x2, 1x3 or 1x4 splitters.
IV.   Split ratios from 1x2 up to 1x32 and higher (dual input possible as well).

V.   Higher split ratios have typically higher IL (Insertion Loss) and lower uniformity compared with planar technology.

- Planar Lightwave Circuit (PLC)

I.   optical paths are buried inside the silica chip available from 1x4 to 1x32 split ratios and higher, dual input possible also
II.   Only symmetrical splitters available as standard devices
III. compact compared with FBT at higher split ratios (no cascading)
IV.   better insertion loss and uniformity at higher wavelengths compared with FBT over all bands
V.   better for longer wavelength, broader spectrum

Common available split ratios:

- 1:2, 1:4, 1:8, 1:16, 1:32, 1:64
- 2:X are required for Type B protection or Dual protection.

2.4.3 **Optical Network Terminal ONT Optical Network Terminals (ONTs)** are deployed at customer's premises. ONTs are connected to the OLT by means of optical fiber and no active elements are present in the link. In GPON the transceiver in the ONT is the physical connection between the customer premises and the central office OLT. WDM triplexes module separates the three wavelengths 1310nm, 1490nm and 1550nm (for CATV service). ONT receives data at 1490nm and sends burst traffic at 1310nm. Analogue video at 1550nm is received. Media Access Controller (MAC) controls the upstream burst mode traffic in an orderly manner and ensures that no collisions occur due to upstream data transmission from different homes. They are fiber to copper media converters that offer RJ11, RJ45, and F-Series connectors to any device. These devices are available in many configurations and port densities up to 24 ports. ONTs are available for outdoor and indoor use, provide POE or no POE, 10/100/1000, AES encryption, and can include batteries for survivability in the event of a power outage. GPON uses Dynamic Bandwidth Allocation that is it dynamically allocates the bandwidth depending on the number of packets available in the T-CONT. Once the OLT reads the number of packets waiting in T-CONT it assigns the bandwidth. If there are no packets waiting in the T-CONT, then OLT assigns the bandwidth to other T-CONT which has packets waiting in T-CONT. If an ONT has a long queue OLT can assign multiple T-CONTS to that ONT

2.5 TRAFFIC FLOW IN GPON FTTH ACCESS NETWORKS

The data is transmitted from OLT to ONT in downstream as a broadcast manner and as a Time Division Multiplexing (TDM) in upstream. The wavelength of the downstream data is 1490 nm, voice and data services from core network transported over the optical network reaches the OLT and are distributed to the ONTs through the FTTH network by means of power splitting. Each Home receives the packets intended to it through its ONT. The upstream represents the data transmission from the ONT to OLT. The wavelength is 1310 nm. If the signals from the different ONTs arrive at the splitter input at the same time and at the same wavelength 1310nm, it results in superposition of different ONT signals when it reaches OLT. Hence TDMA is adopted to avoid the interference of signals from ONTs. In TDMA time slots will be provided to each user on demand for transmission of their packets. At the optical splitter packets arrive in order and they are combined and transmitted to OLT.

Type B protection is used in the design of the GPON FTTH access network presented in this paper. It provides redundancy against both feeder and GPON port failures. In this type of the protection, each fiber strand in the feeder cable is connected to two GPON ports in the OLT. One of the ports is configured as active and the other as standby. the traffic flow from working path in case of failure of port or fiber cut then traffic automatically diverted on protection path and reach up to standby port of OLT.

2.6 DESIGN **VALIDATION**

In order to assess the feasibility of the proposed design of the FTTH network and that each user in the network can receive adequate power , the total optical power loss between the GPON port of the OLT and that of the ONT should be considered. ONT sensitivity is -26 db hence design loss should not be greater than - 26 db and also provide margin to operations team for maintenance activity.

This loss can be summarized by the following equation:

$loss = lcable + lsplitter + lsplice + lconne\ ctor$

- *lcable* : Account for the loss in optical signal power as it traverse the fiber cable, it is measured in dB/Km. Typically design value is 0.3 db/km for 1310 nm and 0.22 db/km for 1550 nm.

- *lsplitter* : Refers to splitter insertion loss, it varies according to the splitting ratio. The values of used for this parameter are obtained from datasheets of the corresponding splitter.

Typically design vale of Level-1 splitter is 10 db and level-2 splitter is 18 db.

- *lsplice* : Represent the loss introduced due to splicing; it is measured by the fusion splice machine. Typically value is 0.01 db

- *lconne ctor* : Manifests the loss introduced by connectors coupling. Typically design value is 0.5 db

Questions

1. What are main part of FTTx ?
2. How many types of splitter?
3. Is splitter is passive device?
4. What is function of OLT?
5. What is link loss formula?
6. Difference between feeder, distribution and drop cable?
7. What is point to multi point connectivty?

CHAPTER-3

Optical Fiber Cable

An **optical fiber cable** is a bundle of fibers that are used to carry light.

Optical fiber (or fiber optic) refers to the medium and the technology associated with the transmission of information as light pulses along a glass or fiber. Optical fiber carries much more information than conventional copper wire and is in general not subject to electromagnetic interference and the need to retransmit signals. Most telephone company used optical fiber cable for long-distance lines. The glass fiber requires more protection within an outer cable than copper.

Optical fiber cable is made up into multiple layers.

- Fiber
- Loose tube
- Strengthen member
- Plastic sheath with metallic tape
- Jelly
- Metallic sheath (only in Armored optical fiber cable)
- cable sheath or outer sheath of fiber

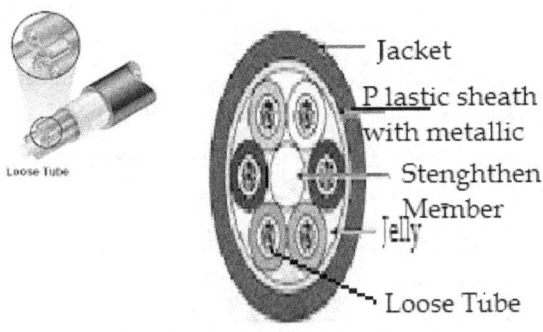

Loose Tube

Jacket

Plastic sheath with metallic

Stenghthen Member

Jelly

Loose Tube

OPTICAL FIBER CABLE

There are mainly two design of optical fiber cable:

- Armored cable
- Unarmored cable

The difference between armored and unarmored cable is only metallic sheath coating is available in armored optical fiber cable. This metallic sheath provide extra protection to optical fiber cable and helpful in identification of fiber fault.

Sometimes dummy loose tube will place in optical fiber cable to proper binding. Optical fiber cable is available in bunch of 12 F, 24 F, 48 F, 96 F etc. and also available in different color e.g. orange, black, green etc.

Where F stands Fiber.

Loose Tube and Color coding of fiber

There are 12 standard colors coding of fiber. And same fiber colors are used for loose tube. Below mention chart reflect the color coding of 48 Fiber.

Loose Tube -4 No. , Fiber-48

	Blue	Orange	Green	Brown	Grey	White	Red	Black	Yellow	Violet	Pink	Aqua
Loose Tube						Fiber No.						
Tube 1	1	2	3	4	5	6	7	8	9	10	11	12
Tube 2	13	14	15	16	17	18	19	20	21	22	23	24
Tube 3	25	26	27	28	29	30	31	32	33	34	35	36
Tube 4	37	38	39	40	41	42	43	44	45	46	47	48

Loose Tube-8, Fiber-48

Loose Tube	Blue	Orange	Green	Brown	Grey	Aqua
				Fiber No.		
Tube 1	1	2	3	4	5	6
Tube 2	7	8	9	10	11	12
Tube 3	13	14	15	16	17	18
Tube 4	19	20	21	22	23	24
Tube 5	25	26	27	28	29	30
Tube 6	31	32	33	34	35	36
Tube 7	37	38	39	40	41	42
Tube 8	43	44	45	46	47	48

Loose Tube-6, Fiber-48

Loose Tube	Blue	Orange	Green	Brown	Grey	White	Red	Aqua
				Fiber No.				
Tube 1	1	2	3	4	5	6	7	8
Tube 2	9	10	11	12	13	14	15	16
Tube 3	17	18	19	20	21	22	23	24
Tube 4	25	26	27	28	29	30	31	32
Tube 5	33	34	35	36	37	38	39	40

Tube 6	41	42	43	44	45	46	47	48

Splicing/Termination of Fiber

Splicing is the process to joint two bare fibers; two techniques are generally used for jointing of fibers.

- Mechanical splicing
- Fusion splicing

When light is propagating through fiber some loss is reflecting at joints which are known as splicing loss.

Mechanical splicing is not commonly used due to high splicing loss and not reliable, hence fusion splicing has become more popular.

Termination is process to joint one end of fiber with pigtail (one end fiber and other end is fiber connector) for terminated on link into FMS.

List of tools used in splicing

- Optical fiber sheath cutter
- Loose tube cutter
- Fiber stripper
- Fiber cleaver

Mechanical Splicing

Mechanical splicing is the process to joint two fibers with the help of mechanical sleeve.

It is used in case of emergency where prepared fiber length is very short or high loss will not affect the fiber link design budget. It is quick

splicing process. **Process of Mechanical Splicing**
Step I: Preparation of fiber cable sheath

a. Expose the rip cord. This step involves ring-cut the outer jacket with a sharp knife or cable sheath cutter, remove the corrugated armor if applicable, and shave off the outer jacket to expose the rip cord.

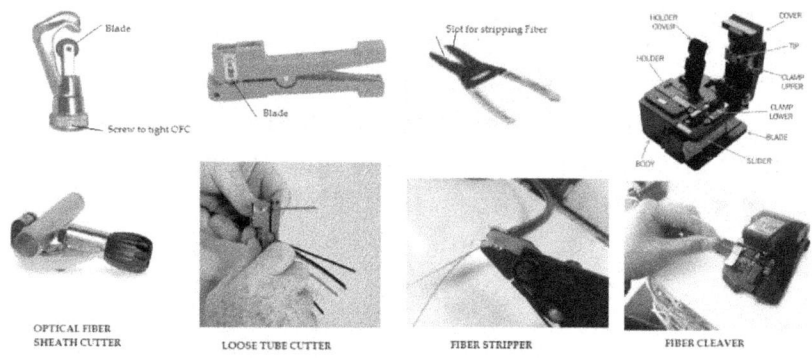

b. Remove the outer sheath. This step involves make a longitudinal cut the outer sheath by rip cord and peel off the outer jacket and corrugated metal.
c. Remove the inner jacket. This step involves cut central strength member, and clean the filling compound with IP solution.

App. 1 to 1.5 meter optical fiber cable is required for splicing and routing of fiber.

Step II: Expose fiber by remove loose tube protection.

Step III Identify number of fiber by color code sheet.

Step-IV Remove color coating of fiber by stripper and cleave the fiber by cleaver.

Step V Insert cleaves fiber at mechanical sleeve.

Now mechanical splicing completed.

Step VI Repeat Step II to Step V for all fibers in optical fiber cable.

Step VII Route all spliced fiber in splicing tray of joint closure.

Fusion Splicing

Fusion splicing is the process of jointing two bared fibers by arc or heat technique. It is introducing very nominal splice loss and more reliable methodology.

It is most popular splicing technique.

Process of Fusion Splicing

Step I: Preparation of fiber cable sheath

a. Expose the rip cord. This step involves ring-cut the outer jacket with a sharp knife or cable sheath cutter, remove the corrugated armor if applicable, and shave off the outer jacket to expose the rip cord.
b. Remove the outer sheath. This step involves make a longitudinal cut the outer sheath by rip

cord and peel off the outer jacket and corrugated metal.
c. Remove the inner jacket. This step involves cut central strength member, and clean the filling compound with IP solution.

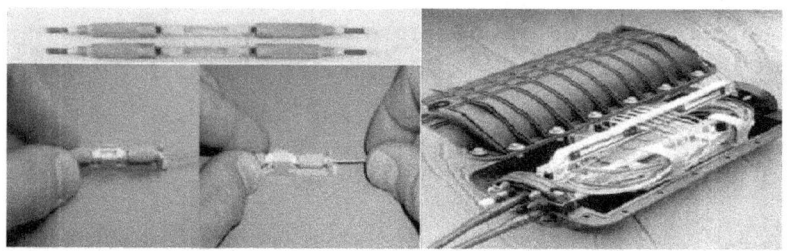

MECHANICAL SLEEVE &
MECHANICAL SPLICING   JOINT CLOSURE

App. 1 to 1.5 meter optical fiber cable is required for splicing and routing of fiber.

Step II: Expose fiber by remove loose tube protection.
Step III Identify number of fiber by color code sheet.
Step IV Insert protection splicing sleeve in one end of fiber.

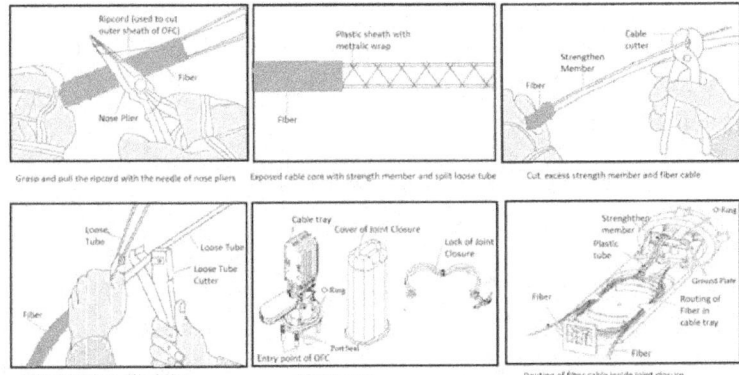

Grasp and pull the ripcord with the needle of nose pliers   Exposed cable core with strength member and split loose tube   Cut excess strength member and fiber cable

Cut loose tube and exposed fiber cladding   Fiber Joint closure   Routing of fiber cable inside joint closure

Step V Remove color coating of fiber by stripper and cleave the fiber by cleaver.

Step VI Insert cleaves fiber in fusion splicing and aligned cleaved fibers.

Step VII Select fusion option which produce sufficient arc in which tip of fibers are melt and form a joint. This is known as fusion splicing.

Step VIII Cover the fusion splicing joint with protection splicing sleeve.

Step IX Insert protection splicing sleeve in heat sink. Heat sink is generating sufficient heat to melt protection sleeve and protection sleeve will fix on fusion spliced fiber as protection cover.

Step X Repeat Step II to Step V for all fibers in optical fiber cable.

Step XI Route all spliced fiber in splicing tray of joint closure.

Joint Closures

It is protection device of fiber joint and it is utilized for storage of Fiber sleeves and routing of optical fiber cable.

Main part of Joint closure

- Outer cover of Joint closure
- O-Ring with lock of Joint closure
- External ground port
- End plate to tight Strengthen member
- Optical fiber cable entry port
- Splicing tray
- Port seal

How to optical fiber cable installed in Joint Closure?

Step-by-step processes of joint closure installation are mention below.

PROCESS OF FUSION SPLICING

Step I Preparation of fiber cable sheath
   a. Expose the rip cord. This step involves ring-cut the outer jacket with a sharp knife or cable sheath cutter, remove the corrugated armor if applicable, and shave off the outer jacket to expose the rip cord.
   b. Remove the outer sheath. This step involves make a longitudinal cut the outer sheath by rip cord and peel off the outer jacket and corrugated metal.
   c. Remove the inner jacket. This step involves cut central strength member, and clean the filling compound with IP solution.
      App. 1 to 1.5 meter optical fiber cable is required for splicing and routing of fiber.

Step II Enter prepared optical fiber cable into joint closure through entry port and insert port seal to each optical fiber cable.

Step III Now unscrew knob and tighten Strengthen member of optical fiber cable. It is fix the optical fiber cable in joint closure.

Step IV Connect armored cable into ground plate. It is used to locate underground OFC by the help of cable locator.

Step V Tie loose tube cable on splicing tray.

Step VI Routing of optical cable into splicing tray.

Step VII Place moisturizer absorbing bag to restrict insect inside joint closure.

Step VIII Lock the joint closure.

Step IX Hot gun blower is used to seal entry cable port of joint closure. Before using hot gun blower wrap aluminum foil into optical fiber cable which protect OFC from damage from heat.

Hot gun blower is equipment which produces hot air.

How to perform drum testing of OFC

Step I: Preparation of fiber cable sheath

a. Expose the rip cord. This step involves ring-cut the outer jacket with a sharp knife or cable sheath cutter, remove the corrugated armor if applicable, and shave off the outer jacket to expose the rip cord.

b. Remove the outer sheath. This step involves make a longitudinal cut the outer sheath by rip cord and peel off the outer jacket and corrugated metal.

c. Remove the inner jacket. This step involves cut central strength member, and clean the filling compound.

Step II: Expose fiber by remove loose tube protection.

Step III Remove color coating of fiber by stripper and cleave the fiber by cleaver. An also strip and cleave the fiber end of pigtail.

Step IV Now put bared fiber and pigtail in fusion splicing machine for alignment of the bared fiber with pigtail. To ensure that pigtail and fiber should not spliced.

Step V Fiber connector end of pigtail is connected with OTDR and launch pulse of light to check health of OFC drum.

Step VI Repeat step II to step V for all fibers in optical fiber cable.

If no bending or damaged loss reported; means OFC is healthy.

For 48 Fiber OFC, this test is repeated by 48 times to check health status of all fiber.

Link Budget Fiber Loss calculation formula

Link budget Fiber Loss means total fiber loss occurred in fiber link (from **X** location to **Y** location). All fiber link planning are based on link budget fiber loss.

Link Loss $= F_L \times A_W + S_N \times 0.1 + 2 \times C_L$

F_L -- Fiber Length in Km

A_W -- Attenuation Loss per Km.

Attenuation Loss is depending on wavelength used.

1310 nm – 0.35 dbm

1550 nm – 0.22 dbm

S_N – Number of Splicing

C_L – Connector Loss (Generally consider 0.5)

Let us explain with example, assuming

- Optical fiber length -- 40 km
- no. of splice -- 10
- Wavelength – 1310 nm
  Then, Link Loss $= 40 \times 0.35 + 10 \times 0.1 + 2 \times 0.5$

  $= 16$ dbm

- Optical fiber length -- 40 km
- no. of splice -- 10

- Wavelength – 1550 nm
  Then, Link Loss $= 40 \times 0.22 + 10 \times 0.1 + 2 \times 0.5$

  $= 10.8$ dbm

Difference between db and dbm

A dB is a RELATIVE measure of two different POWER levels.

For amplifiers, a common reference unit is the dbm, with 0dBm being equal to 1 milliwatt

FMS Module

Fiber Management System module is basically termination point of optical fiber cable. In the FMS fiber connector is installed.

Fiber connector

Fiber connector is interface between optical fiber cable and Fiber optics equipment.
List of commonly used fiber connector are mention below:

- FC Connector
- ST Connector
- SC Connector
- LC Connector
- MU Connector
- E2000 Connector

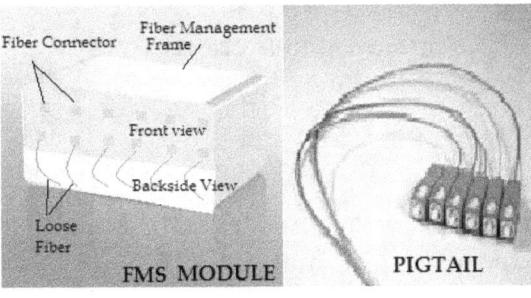

FMS MODULE   PIGTAIL

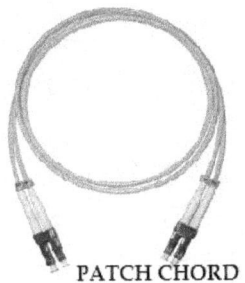

PATCH CHORD

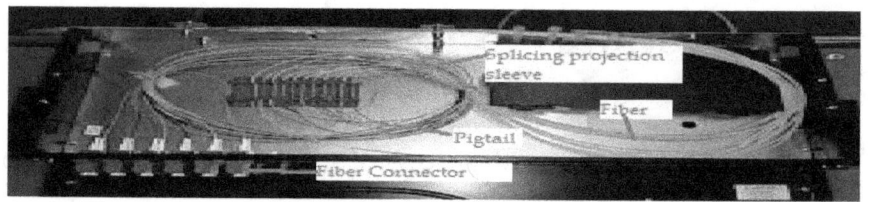

Pigtail

Pigtail is made by optical fiber cable where one end is optical fiber cable and other end is optical fiber connector. It is used to terminate optical fiber cable in FMS.

Patch Chord

Patch chord is made by optical fiber cable where both ends are fiber connector. It is basically used to connect FMS and Fiber optic equipment.

Proper labeling has to be done on patch chord for identification in future.

Labeling

Labeling is very important part of every activity. It helps to identify right material in future use.

Some common place of labeling and purpose are mention below:

- Manhole/Hand-hole – Labeling has to be performed on optical fiber cable to identify direction of optical fiber cable in future.

- Joint Closure – Labeling will be done on below mention materials :
  a. Splicing tray – To locate number of fiber in Joint closure.
  b. Loose tube – To locate loose tube and optical fiber cable direction.
  c. Optical fiber cable – To locate direction of optical fiber cable.

- Patch Chord- To locate right patch chord in future.
- Pigtail – To identity pigtail is connected with which port and optical fiber cable.
- FMS – To identify number of fiber port and also direction of optical fiber cable.

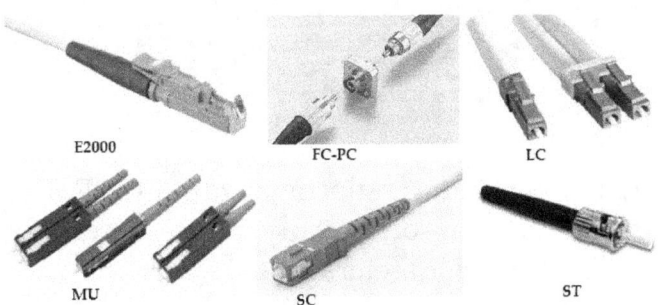

E2000   FC-PC   LC

MU   SC   ST

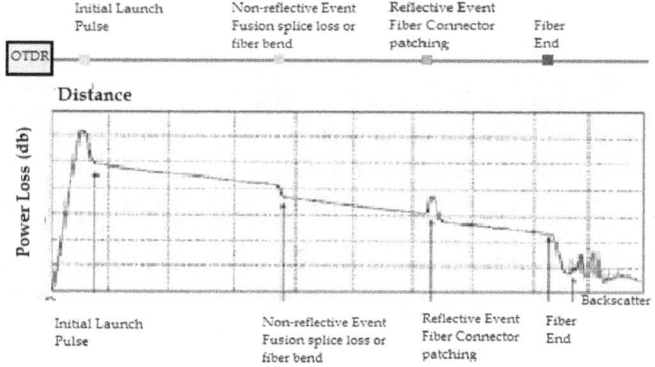

OTDR TRACE REPORT

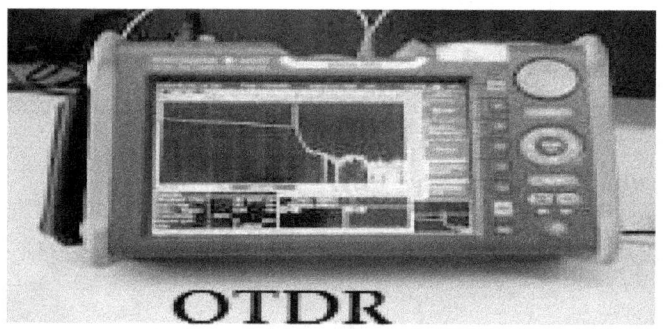

OTDR

Any fiber bend reflecting into OTDR trace then it need to rectify before Laser source & power meter testing.

- Capture Total Link Loss by using Laser source & power meter.

Reading of laser source and power meter testing (LSPM) should be less than total link designed budget.

Some probable cause of high LSPM reading
- High fiber splicing loss.
- Fiber bend
- Faulty fiber connector.

How to perform LSPM testing

- **Step –I** : Measure loss of patch chord used in Laser source-power meter test.

As shown in the figure connect two patch chord one with laser source and second patch chord on power meter and generate pulse of predefined wavelength from laser source and Capture power reading at power meter .

E.g. laser source generate +5 dbm pulse of 1550 nm and power meter receives +3 dbm powers. It means loss of patch chord is 2 dbm.

- **Step-II**: Connect laser source at one end of fiber and power meter at other end.

Generate pulse of predefined wavelength from laser source and Capture power reading at power meter. Subtract loss of patch chord and fiber connector loss from measured power reading the final reading is known as total link loss of optical fiber cable.

E.g. Laser source generate +5 dbm pulse of 1550 m and power received reading -20 dbm

Then, Total link loss = - 20 - (patch chord loss) – 2 x (fiber connector loss)

= - (20- 2- 2x0.5) = -17 dbm

What is ON-OFF-ON technique used in LSPM?

Sometimes power meter connect on wrong fiber number and consistently receives power level to overcome this problem there are three basis steps

Step –I First ON laser source and receive power level.

Step-II Switch off laser source and verify that no power level at power meter.

Step-III Again switched on laser source and received power level at power meter. That confirms laser source and power meter connected on right fiber.

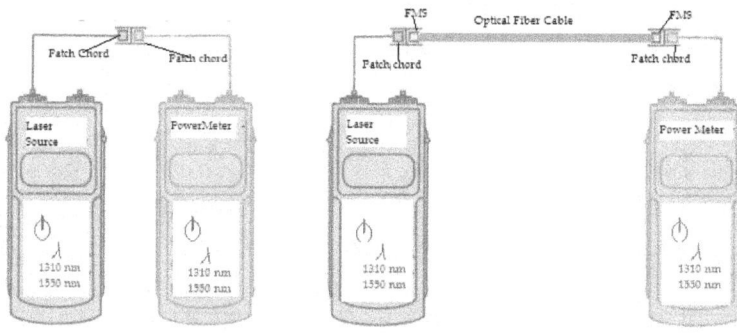

Measure loss of Patch chord used in LSPM

Measure loss of Optical fiber cable

Laser Source & Power meter

testing

LINK	A TO B		LINK	B TO A	
Fiber No.	1310 nm	1550 nm	Fiber No.	1310 nm	1550 nm
1			1		
2			2		
3			3		
4			4		
5			5		
6			6		

Questions

8. What are parts of optical fiber cable?
9. What are color sequences of optical fiber cable?
10. How many loose tubes are in optical fiber cable?
11. What is means by dummy loose tube?
12. What is use of strengthen member in optical fiber cable?
13. What is mechanical splicing?
14. What is fusion splicing?
15. What is use of laser source and power meter?
16. How LSPM testing performed?
17. What is OTDR?
18. What is meaning of reflective event and non-reflective index?
19. How to calculate optical fiber link budget?
20. What is patch chord?
21. What is pigtail?
22. What is Fiber connector and types of fiber connector?
23. Describe joint closure and its part?
24. What is FMS module?

CHAPTER-4

Fiber Project Execution

Once route survey has completed, fiber laying plan is ready for fiber project execution. It means excavation work can be started, but before starting excavation we need to reconfirm the RoW authorities and RoW permission.

An **excavation** is any man or machine -made cut, tunnel or trench in an earth surface formed by earth removal.

Trenching, Ducting and Backfilling

Trenching

Trenching is excavation process refers to a narrow excavation made below the surface of the ground in which the depth is greater than the width. Generally it is 1.5m X 1 m (W X D) from NGL. Excavation can be done by machine or manual labor. The aim is to achieve defined depth possibly in straight line.

To achieve straight trench, use limestone marking before starting excavation work. Flat and uniform depth of trench should be maintained and should avoid sharp changes in depth because it can damage or puncture the duct during ducting process.

The depth of trench will be measured from NGL.

NGL mean natural ground level. Always consider the lower side of the ground level.

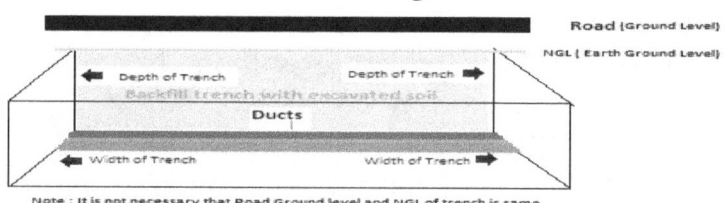

Note : It is not necessary that Road Ground level and NGL of trench is same.

Ducting

Duct or **HDPE** pipe is hollow pipe made up of plastic material where optical fiber cable will be blown into the hollow space. Life of duct is 20 years in underground ambience. Generally the outer diameter of duct is 40 mm and inner diameter is 33 mm.

Ducting is the process where duct has to be laid in the trench. As per organization requirement one or more duct can be laid in the same trench. If more than one duct is laid in same trench, then use different colors of duct which helps easily to differentiate the ducts for future use or maintenance of fiber.

Ducts will be laid in a flat bottom trench, free from stones, and sharp edged debris. The duct will be placed in trench as straight as possible, however at bends, horizontal and vertical minimum bending radius for duct should be maintained.

The ducts will be joined with couplers by using duct cutter & other tools and will be tightened and secured properly. The duct joint should be airtight to ensure smooth cable blowing by using cable blowing machines.

When ducting is completed always use end plug of duct to prevent it from mud, water and dust. Use duct decoiler to decoil the duct, it prevents duct from bend, twist etc.

Duct coupler-is being used to join two ducts.

Duct End Plug – It is cap of duct. It is being used to prevent it from mud, water and dust.

Duct Simplex plug – It is generally used in operation of optical fiber cable. It is being used to join ducts when OFC is already laid in duct.

| DUCT | DUCT COUPLER | DUCT SIMPLEX PLUG | END PLUG | COUPLER |

Backfilling

Backfilling is the process to fill the trench by excavated soil and by ramming to bring the trench at original ground level. During backfilling, avoid direct contact of stones, bricks to the duct; it can damage/puncture the duct.

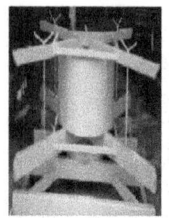   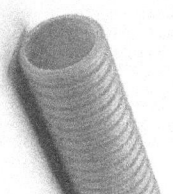

DUCT DECOILER   DUCT   DUCT BUNDLE   DWC PIPE

Methodology of Trenching, Ducting and Backfilling

There are mainly two types of trenching technology

- Open trenching
- Trenchless

Open trenching

It is a process to execute trenching, ducting and backfilling of trench by an open cut method.

Open cut method is a process of cutting the upper part of ground either by machine or manual digging method. The most popular open trench machine is JCB.

Comparison between manual digging and machine trenching

Parameter	Manual Digging	Machine Digging
Cost	Cheaper	Slightly higher
Productivity	Slow	Higher

		Requires proper space for machine movement
Flexibility	Requires less working space	
Quality	Trench is not straight.	Trench is straight as much as possible and no depth issue.
Reliability	Medium	Higher

In rocky terrain, when manual labor and excavation machine is not able to achieve proper depth of trench, then they use rock breaker machine which is specially designed to work in rocky areas. When working in hard rocky area, sometimes rock breaker machine is also ineffective then they use blasting methodology to achieve proper depth. But blasting methodology is not used in every location.

OPEN TRENCHING MACHINE-JCB   TRENCH   BUCKET   ROCK BREAKER

Trenchless

Trenchless technology is basically making a tunnel below the ground without affecting the upper part of the ground.

In trenchless technology Trenching and ducting are performed simultaneously. It is very successful technology for crossing the utilities, rivers, gas pipelines etc. But it is costly as compared to an open trenching and also having some limitations.

The most popular trenchless machine is HDD (Horizontal Drilling Direction).

It cannot work in rocky areas, sandy soil and cannot be used for less than 50 m length.

There is one more trenchless technology is moiling. In this technology tunnel is being done by manual drilling method it is used only for road crossing, only if the road width is less than 20 m.

Type of Soil

Soil is a mixture of sand, silts, clay, water and air Soil is the main factor to be considered for excavation methodology.

Methodology of excavation will directly affect the cost of project.

- Normal Soil – Both
- Hard Soil – Both
- Soft Rock – Open Trenching and Rock Breaker

- Hard Rock – Open Trenching, Rock Breaker and Blasting.
  (Note: Blasting permission is required before excavation from RoW Authorities and should be executed with proper safety plan.)

Protection Technique

Incase depth cannot be achieved due to any reason e.g. other utility, rocky area, digging not possible due to water coming into trench etc. then use protection technique which provides additional protection to duct. Protection technique is additional cost of excavation activity; hence it is used when there is no other option to achieve proper depth.

Below mention protection will be used in case of improper depth.
- DWC Pipe
- PCC refer Cement wall has created around duct.
- RCC Slab
- GI Pipe

Methodology for crossing

- **Minor Road crossing** –We can use open trenching technology and provide GI/DWC pipe support above to duct.
- **Major Road Crossing**—we can use moiling technique or HDD technique.

- **River/Bridge crossing** – We can use HDD or Clamping technique along parapet of bridge. As per the permission provided by RoW Authorities.
- **Railway Crossing** – As per the permission provided by Railway authority.
- **Gas pipeline** –Both excavation technologies can be used. As per the permission provided by Gas authority.

Manhole/Hand-hole

Manhole is a maintenance chamber in which spliced fiber and fiber loop is being placed. We have to ensure that proper space should be available for storage of fiber slag, joint closure installation and enough working space for movement of fiber optics expert inside the manhole.

In manhole both end ducts are terminated and PCC has to be performed at the bottom of the chamber to restrict the entry of mud and water inside the manhole. The manhole will be filled with sand up to manhole cover.

Joint closure is the equipment used to offer room to fuse splice optical fiber and also to provide protections for the fused fiber joint point and the fiber cables.

There are different size of joint closure are available for 24 Fiber Joint, 48 Fiber Joint, 96 Fiber Joint.

Clamps to hold GI pipe

GI pipe (inside GI pipe duct will laid)
Duct

CROSSING OF BRIDGE BY CLAMPING METHODOLOGY

Hand-hole is only used for fiber slag storage and keeping in mind that it will easily converted into manhole in future.

The shape and size of manhole/hand-hole will change as per requirement. Generally pre-casted RCC manhole is used, but sometimes it can be casted on field when sufficient space is not available to install manhole (pre-casted, RCC)

DIT (Duct Integration Testing)

The purpose of Duct integration testing is to verify that the installed ducts or pipes are ready for blowing of the fiber and ducts are through from one manhole to other manhole. Some reasons for lack of duct continuity are:

- Duct damaged
- Couplers not connected

List of duct test are mention below:

✓ Air Test
✓ Shuttle Test
✓ Sponge Blowing Test
✓ Pressure Test

Air Test

Air test is performed to check the duct continuity.

How to Perform Air Test

Air is passed into the duct from one end of hand-hole and check at the other end of hand-hole. Air Compressor is used to generate air in the duct by the help of pressure gauge and lever will control air pressure flow. To ensure that proper calibrated gauge is used.

- If received Normal flow of air at the other end, it means Air test is pass and moved to next DIT test.
- No flow of air or less than normal flow of air at the other end of the duct, it means Air test is failed.

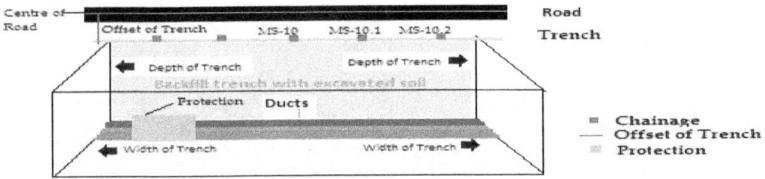

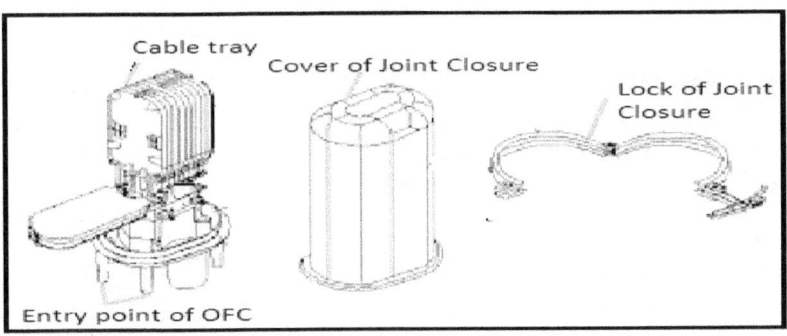

Fiber Joint closure

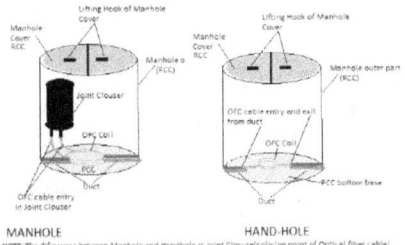

Identify causes of Air test failed and arrest the fault. Some probable causes are mention below:

Kink

- Duct bends
- Ducts damaged during backfilling by sharp edges stone or other third party excavation work.

Blockage
- Coupler leakage.
- Coupler damaged.
- Duct puncture/damaged
- Coupler not connected.

Shuttle Test

Shuttle test is process to identify duct are free from bend or damaged. It is also helpful to identify the exact location of fault.

How to perform Shuttle Test

Shuttle is passed in duct through air compressor machine or DIT machine from one end of hand-hole. If shuttle reached at the other end of hand-hole, then shuttle test is passed.

The size of shuttle should be is 80% diameter of duct and length of shuttle should be 150 mm.

If shuttle not reached at other end of hand-hole, then identify the exact fault location and rectify the fault and repeat shuttle test before proceeding for next DIT test.

How to identify fault location

The shuttle will get stuck at the point where the duct is either kinked or blocked.

Transmitter is passed through the duct to identify the exact location of fault. It gets stuck behind the shuttle. The path of the duct is then tracked with a receiver. As the receiver passes over the transmitter, the signal from the transmitter is heard loud and clear. The spot is marked and exposed duct by digging the trench. Remove the cause of kink or blockage and replaced the said portion with good duct.

Sponge blowing test

Purpose of Sponge blowing test to clean duct from mud, water particle etc.

How to perform of Sponge Blowing test

Sponge is blown to duct by help of air compressor or DIT machine. The diameter of sponge is twice the diameter of duct and 100mm in length

If sponge reach at other end of duct it means duct are cleaned from mud and water particle.

Pressure Test

Pressure test is last test of DIT and very important test. If this test passes then duct ready and fit for fiber blowing.

How to perform of Pressure test

The pressure test is conducted at 5 bar pressure in duct and over a 30 minutes period the pressure loss should not exceed 0.5 bars, then Pressure test passed.

Reason of failure of pressure test

- Leakage at Couplers

- Puncture in Duct

Optical Fiber Cable Blowing

Once DIT test is passed duct are ready for the blowing of fiber. The process of optical fiber cable pulling in duct is called blowing.

Before blowing of fiber, we need to perform OFC (Optical Fiber Cable) drum testing and check the health of optical fiber cable.

Type of blowing methodology
- Manual Fiber Blowing
- Fiber Blowing through Blowing Machine

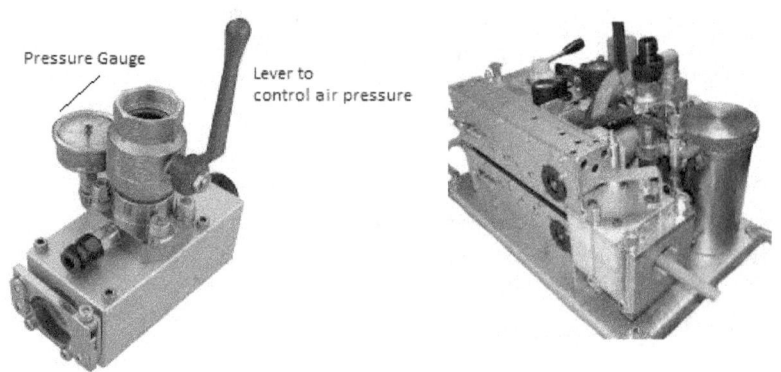

Pressure Gauge

Lever to control air pressure

DIT Machine /Blowing Machine

Manual Blowing of Fiber

Fiber cable blowing will be done by cable rodder. This technique is very useful for short distance cable blowing.

Cable rodder is small size cable coated with steel tape and PE sheath and coil in stand. It is light weight, hard and highly flexible.

How to perform Fiber Blowing via cable rodder

Cable rodder passes through duct and pull the rodder till it will not reached to other end of hand-hole, once rodder reach at other end of hand-hole, then optical fiber cable will tie with cable rodder and pull back rodder with OFC.

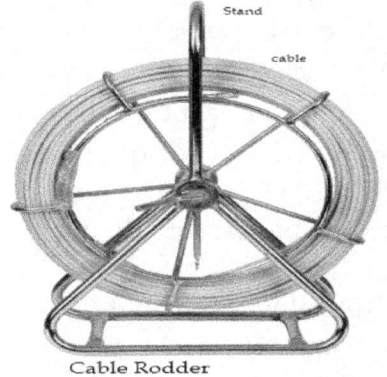

Cable Rodder

Fiber blowing through blowing machine

Optical fiber cable will be blown by the help of blowing machine.

How to perform Fiber Blowing via blowing machine

First of all, uncoiling the OFC drum and make eight structure for free movement of OFC during blowing.

A properly sized cable carrier is attached to the end of the fiber for smoothly and tension free movement of fiber inside the duct. Before inserting the carrier into the duct, cable blowing lubricant is added to reduce the friction.

Cable blower is connected to the system; compressed air or blowing machine is injected into the duct behind the cable carrier. And generate air pressure for fiber blowing due to high air pressure fiber cable move through the duct but avoid high pressure it will produce more friction. Once it reach other side of hand-hole optical fiber cable will be blown into duct.

How to slag a cable in Manhole/Hand-hole

Optical fiber cable is very sensitive and requires more handling with care. It should be ensure that minimum bend radius should not be more than 45° Degree and cable should be tension free. Now coil the OFC in manhole and tie the coil.

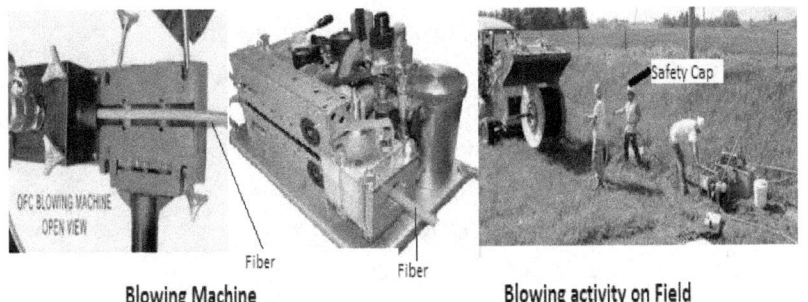

Blowing Machine   Blowing activity on Field

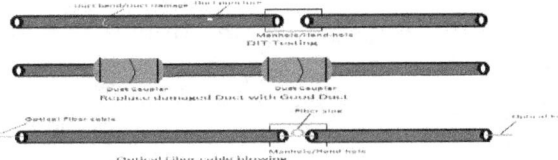

Questions

1. What is meaning of excavation?
2. What are trenching, ducting and backfill?
3. Why compression is necessary in backfilling?
   Hint – If compression not done properly in backfilling, then cavity reflecting on trench and any person or animal can injured on trench.
4. What is NGL?
5. What is outer and inner diameter of duct?
6. What precaution will take during ducting?

Hint- Ducts will be laid in a flat bottom trench, free from stones, and sharp edged debris. The duct would be placed in trench as straight as possible, however at bends horizontal and vertical minimum bending radius for duct would be maintained.

7. What are protection techniques in case of low depth?
8. What are methodologies of excavation?
9. What is limitation of HDD machine?
10. Why Rock breaker is used?
11. What is use of duct decoiler?
12. What is manhole and hand-hole?
13. What safety precaution should be taken during excavation?
14. What is use of end plug?
15. Why PCC has to be done at bottom of Manhole?
16. Which machine is used for open trenching?
17. What to cross underground utilities by open trenching?
    Hint-
    A. Expose underground utilities and then tries to make depth without damaged utility.
    B. Divert the trench.
    C. Use protection technique.
18. What is rod length of HDD machine?

Hint – Generally every rod length is 3 M and some machine having rod length of 4.5 M.
19. What are tests are conducting in DIT?
20. What is pressure level in DIT?

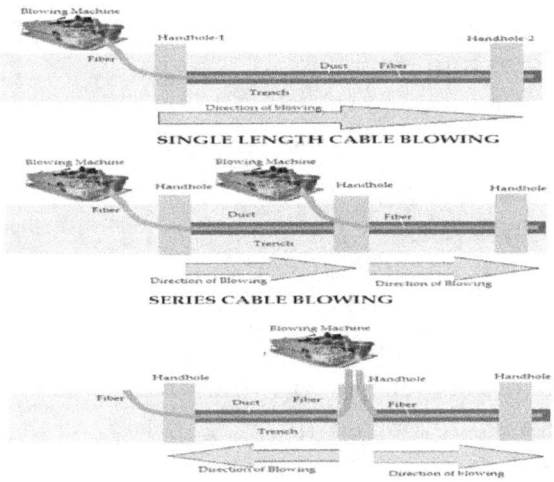

SINGLE LENGTH CABLE BLOWING

SERIES CABLE BLOWING

MID-POINT CABLE BLOWING

CHAPTER-5

Safety rules and regulation

Safety is always prime focus on every project. There are many potential hazards when working in excavations and trenches.

Some safety rules follow during project:

- Excavation area should be covered with warning tape and barrier.
- Wear safety cap during work.
- Wear safety boot and life jacket during work.
- Excavation projects should be carefully planned and implemented to minimize potential damage to buried utilities and to mitigate hazard exposure to personnel.
- Do not work in excavations where water is accumulating unless the water is being controlled and maintained at a safe level
- Provide sufficient lighting to facilitate safe operations at each work location
- Keep heavy equipment away from trench edges and Identify other sources that might affect trench stability, it will avoid cave-in occurs.

- During Excavation ensure that slope of trench no steeper than 45 D.
- Take more precaution during road crossing.
- Never directly look at an optical fiber end or connector end face with the naked eye and do not point at other people. Do not attempt to view at a distance of less than 100mm.
- All fiber ends and waste lengths of optical fiber cable should be disposing carefully because fiber off-cuts are a serious health hazard and once under the skin they are virtually impossible to locate and remove.

SAFETY CAP   SAFETY SHOES   WARNING TAPE AND BARRIER

Fire

Fire is a process in which substances combine chemically with oxygen from the air and typically give out bright light and heat.

Type of Fire & Fire extinguisher

There are five types of fire class and each type of fire has different fire extinguisher.

A **fire extinguisher** is an active fire protection device used to extinguish or control small fires, often in emergency situations only. Typically, a fire extinguisher consists of a hand-held cylindrical pressure vessel containing an agent which can be discharged to extinguish a fire.

It is for the purpose of out-of-control fire, it is suggested that call fire department immediately when fire is out-of-control.

- A class : Ordinary solid combustibles
  e.g. wood
  Fire extinguisher: solid form= Water type
- B class : Flammable liquids and gases
  Fire extinguisher: Foam or Carbon Dioxide (CO_2) or Dry Chemical Powder (DCP) type
- C class: Energized electrical equipment.
  Fire extinguisher: Carbon Dioxide (CO_2), Dry Chemical Powder (DCP) type.
- D class : Combustible metals ;
  Fire extinguisher: Dry Chemical Powder type
- K class: Cooking oil and fats.
  Fire extinguisher: water type.

CHAPTER-6

Machine and Tool Used in FTTx

List of Machine
- ✓ Horizontal Drilling Machine
- ✓ Open-trenching Machine e.g. JCB, Rock Breaker
- ✓ Duct Integration Testing & Blowing Machine
- ✓ Splicing Machine

List of Tool
- ✓ Laser Source
- ✓ Power Meter
- ✓ Optical Time Domain Reflectometer(OTDR)
- ✓ Optical Detector
- ✓ Optical Fiber Cable Locator
- ✓ Global Positioning System(GPS)
- ✓ Duct Rodder
- ✓ Rodometer
- ✓ Duct Cutter
- ✓ Optical fiber sheath cutter

- ✓ Cleaver
- ✓ Stripper
- ✓ Loose tube cutter
- ✓ Dummy fiber spooler

Horizontal Drilling Machine (HDD)

Trenchless technology is basically making tunnel below the ground surface without affect the upper part of ground.

Site inspection and survey is required before using HDD machine to identify underground utility and prevent the same by changing depth of tunnel.

It is very useful technique for crossing of river, major road, gas pipeline etc.

Main part of HDD machine

- Pilot Hole – It drill underground ground surface.

- Mud Motor – To control depth and direction of trench.

- Sonde or transmitter & Hand-held receiver – To monitor depth and direction of trench

- Wash over pipe – Attached with pilot so that soil will not take it original shape after drilling.

- Reamer – use to increase the size of trench hole.

How to locating and pre-determine trench

Sonde or transmitter is installed behind the pilot rod now angle, rotation, direction of pilot rod is encoded into electromagnetic signal and that electromagnetic signal is decoded by hand-held receiver at surface of ground; hence HDD operator can easily monitor the direction and depth of trench.

Working principle of HDD involves three stages

Pre identify entry and exit pit of HDD shot which is known as pre-determine path of trench Now, with help of sonde and handheld receiver easily trench will be done on pre-determine path.

- **Pilot Drilling** – In this process pilot is drilling underground surface and generating small hole on pre-determine path by jet bit.

   Steering element is installed to control the desired drilling direction. Wash-over pipe is attached with pilot so that soil will not fill the hole created by pilot.

- **Pre-reaming** – After arrival of pilot at other side, tunnel has been prepared then pilot is pulled back and reamer is attached with wash-over pipe. The purpose of reamer is to enlarge the hole created by pilot so that duct can be easily pullback.

- **Pull-back pipe** – Duct is attached with reamer and pullback the wash-over pipe.

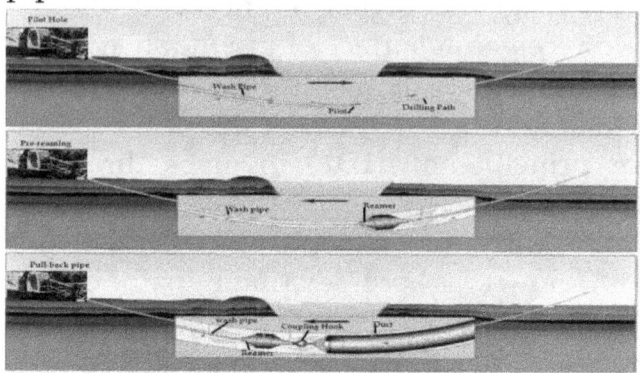

Open trenching Machine

It is process to execute trenching, ducting and backfilling of trench by open cut method. Open cut method is process of cutting the upper part of ground either by machine or manual digging method.

JCB is most common open trenching machine.

Main part of JCB

- Digging Bucket – Used to carry excavated material e.g. soil, rock etc.
- Hydraulic auxiliary – To control depth and direction of Bucket.

- Four wheel drive (including, engine, brakes) – For movement of machine.

Working Principle

Movement of JCB machine by four wheel drive and reached to excavation location, now with help of digging bucket excavation is performed and hydraulic auxiliary is attached to control the direction and depth of trench.

DIT and Blowing Machine

Fiber blowing machine is used to perform DIT test and blown optical fiber cable into duct.

Main part of DIT and blowing machine

- **Aluminum split coupler** – To hold duct and OFC cable so that controlled air pressure   passed to duct without damaging to OFC.
- **Pressure gauge** – To measure air pressure
    Level.

OPEN TRENCHING MACHINE-JCB   TRENCH   BUCKET   ROCK BREAKER

- **Air compressor** – To generate air pressure.

Working Principle

Step-I Connect Duct or OFC to Aluminum Split Couplers (used with Optical Fiber blowing operations to connect the underground laid OFC duct to couple with cable blowing machine at ground level or at convenient locations ,or at when the regular coupling joints may burst during the course of blowing) and

Step-II Add lubricant into duct than generate air pressures to reduce the friction.

Step-III generate air pressure which is controlled by lever at pressure gauge, when sufficient pressure is generated OFC cable start moving inside the duct.

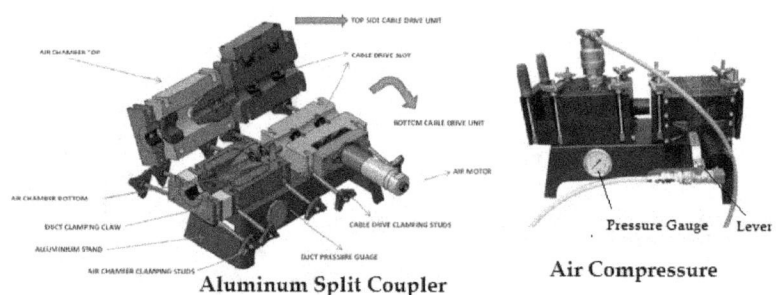

Aluminum Split Coupler

Air Compressure

Splicing machine

Fiber optic fusion splicer is equipment used to joint two bare optical fiber cables. Before using the fiber optic fusion splicer, we need to cut the fiber optic cable and take away all the fiber cable jacket, then use fiber optic cleaver to make the fiberglass end face ready, after finishing these work we can use the fiber fusion splicer to melt the two fiberglass together. This fusion splicer is an automated fusion-splicing machines and it can automatically perform the fusion work after you put the two fibers ready in the machine.

Main part of Splicing Machine
- X-Y motor –To align fiber core
- Electrode – To produce arc for melting of fiber
- Mirror – To monitor alignment and splicing of fiber.
- Fiber holder v-groove – To hold cleaved fiber
- Heat Sink – To provide sufficient heat to protection sleeve so that protection sleeve will encapsulate splice fiber portion.
- LCD screen – To display splicing process

Working Principle of fusion splicing

Before the optical fibers can be successfully fusion-spliced, they need to be carefully stripped of their outer jackets and thoroughly cleaned, and then precisely cleaved to form smooth, perpendicular end faces. Once all of these completed, each fiber is placed into a holder of fusion splicer.

There are three steps involved in fusion splicing:

Step –I Alignment

In this process fiber's position are properly aligned by precise motor. Purpose of alignment of fibers, the light wave should properly propagate through fiber cable resulting low splicing loss. The alignment of fiber are reflects in the LCD screen of fusion machine.

Step-II Impurity Burn-Off

Sometimes traces of dust or other impurities are present on the tip of cleaved fiber; hence prior to fusing, they generate a small spark between the fiber ends to burn off any remaining dust.

Step-III Fusion

After the fibers have been properly aligned and any dusts have been burned off, the electrode generate spark that melts the optical fiber end faces. When melted tips are joined together it means permanent fusion splicing done. A typical splice loss should be 0.1 dB or less which is reflecting in the fusion splicing machine.

Some common factor that will increase splicing loss during fusion splicing

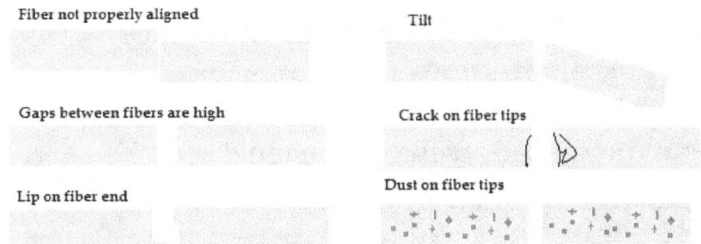

Fiber not properly aligned

Tilt

Gaps between fibers are high

Crack on fiber tips

Lip on fiber end

Dust on fiber tips

FACTORS THAT GENERATE HIGH SPLICING LOSS

TOOLS USED IN FIBER OPTICS

Laser Source
Fiber optic light source is a fiber optic test equipment to measure the fiber optical loss of

71

optical link. It is provide continuous wave of light source e.g. 1310 nm/1550 nm.

Main part of Laser Source
- Laser diode – To generate pulse of light
- Fiber connector – Interface to connect with patch chord.

Working principle of Laser Source
Laser diode is generating continuous source of light with same intensity and also having capability to produce different wavelength.

Power meter
Optical power meter is a test instrument used for absolute optical fiber power measurement. Main parts
- Photo diode – To receive light signal
- LCD screen – Reflect power level
- Fiber connector – Interface to connect with patch chord.

Working principle of power meter

Photo diode receive the modulated light streams and measure the intensity or power level of light signal.

How to perform LSPM testing

Step -I : Measure loss of patch chord used in Laser source-power meter test.

As shown in the figure connect two patch chord one with laser source and second patch chord on power meter and generate pulse of predefined wavelength from laser source and Capture power reading at power meter .

E.g. laser source generate +5 dbm pulse of 1550 nm and power meter receives +3 dbm powers.

It means loss of patch chord is 2 dbm.

Step-II: Connect laser source at one end of fiber and power meter at other end.

Generate pulse of predefined wavelength from laser source and Capture power reading at power meter. Subtract loss of patch chord and fiber connector loss from measured power reading the final reading is known as total link loss of optical fiber cable.

E.g. Laser source generate +5 dbm pulse of 1550 m and power received reading -20 dbm

Then, Total link loss = - 20 - (patch chord loss) – 2 x (fiber connector loss)

$$= - (20- 2- 2x0.5) = -17 \text{ dbm}$$

OTDR (Optical Time-Domain Reflectometer)

OTDR (Optical time-domain reflectometer) is important test equipment used in fiber optic work. During its working process, the optical time domain reflectometer sends a pulse into the fiber cable, the fiber signal light will be scattered back and reflected back because of the fiber bend, fiber joint point or fiber break. The purpose of an OTDR is to detect, locate, and measure events at any location on fiber link.

Main part of OTDR machine
- Laser diode – Generate pulse of light source.
- LCD screen – Used to display OTDR trace
- Photo diode – To measure reflective and non-reflective loss.
- Fiber connector – Used to interface between optical fiber cable
- Input Power supply socket
- Internal Memory - For storage OTDR trace.
- USB port – To export OTDR trace

Input Parameter
- Pulse width – 2ns to 20 micro second (power of light)
- Wavelength – 1310 to 1550 nm
- Range- Distance of optical fiber length
- Averaging- 10 second to 3 minute(for accuracy of events)

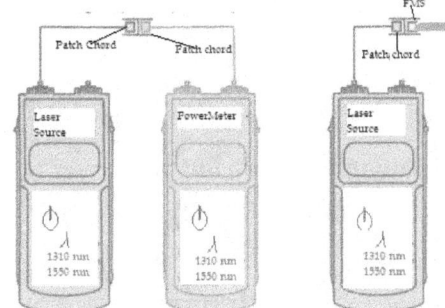
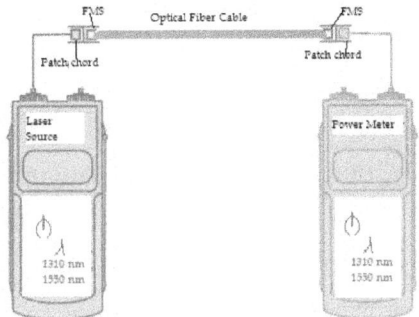

Measure loss of Patch chord used in LSPM

Measure loss of Optical fiber cable

Output parameter

- **Reflective Events** – Reflective events occur where discontinuity arise in the fiber, causing an abrupt change in the refractive index. Reflective events can occur at breaks, connector junctions, mechanical splices or the indeterminate end of fiber.

- **Non reflective events** – Non reflective events occur where discontinuities are absent in the fiber and are generally produced by fusion splices or bending losses.

Working Principle of OTDR: fiber is based on detecting small signals that are returned back to it in response to the injection of a large

75

signal. In this regard, the OTDR depends on two types of optical phenomena: Raleigh scattering and Fresnel reflection.

- **Rayleigh scattering** is intrinsic to the fiber material itself and is present along the entire length of fiber. Rayleigh scattering is uniform along the length of the fiber; therefore, its discontinuities can be used to identify anomalies in the transmission along the fiber link.
- **Fresnel reflection**, on the other hand, is point events and occurs only where the fiber comes in contact with air or another media, such as a mechanical connection, splice or joint.
- **Dead zone** of an OTDR is the distance (or time) where the OTDR cannot detect or precisely localize any event on the fiber link.

How to read OTDR trace report?
Optical detector
It is optical device that used to detect live fiber from bunch of fiber cable.

Main part of Optical detector
- **Photo diode**
- **LED**

- **Slot to hold loose fiber**

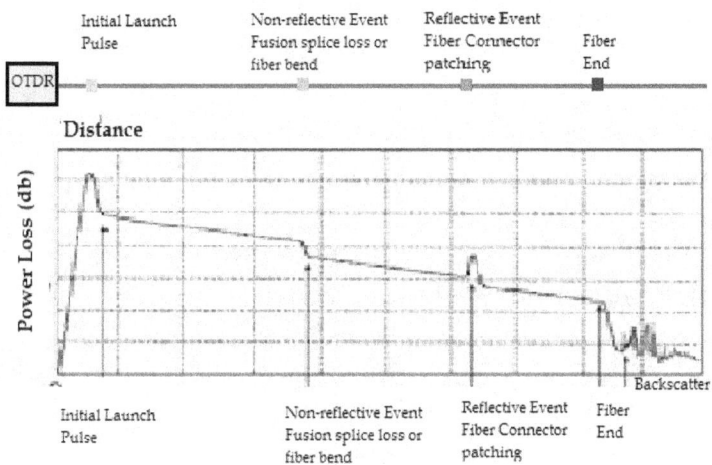

OTDR TRACE REPORT

Working principle
It receive light signal from any source then LED is glow, but it will not reflect the power level of light source.

Optical fiber cable locator
Optical fiber cable locator is device that used to identify underground depth of optical fiber cable and it reflects the underground laid direction of optical fiber cable.

Main part of optical fiber locator
- Transmitter
- Receiver with LCD screen

Working principle
Transmitter is generates electromagnetic signal in metallic sheath of optical fiber cable and electromagnetic signal received by hand-held receiver on the ground surface; hence the depth and direction of optical fiber cable will traced.

GPS (Global Positioning system)
Global Positioning System (GPS) is a space-based satellite navigation system that provides location and time information. Every location on earth have different GPS; hence any location any be identified by GPS reading.

Duct Rodder
Cable rodder is small size cable coated with steel tape and PE sheath and coil in stand. It is light weight, hard and highly flexible.

Rodometer
Rodometer is tool in which one rod is attached with wheel, when there is any movement in wheel it reflects the distant travel by wheel in meter. It is used for measurement of trench, route length etc.

Duct Cutter

Duct cutter is tool to cut the whole duct in uniform manner.

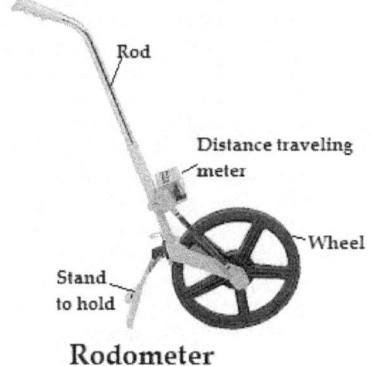

Rodometer

Optical fiber sheath cutter
Optical fiber sheath cutter is tool which is used strip outer sheath of optical fiber cable.

Cleaver
Cleaver is used to cut bared optical fiber before splicing of fiber.

Fiber Stripper
Fiber stripper is tool which used to remove coating of fiber.

Duct Cutter   Dummy fiber spooler   GPS

79

Loose Tube cutter

Loose tube cutter is tool to cut loose tube cable of optical fiber.

Dummy Fiber Spooler

Optical fiber is wrap into spooler which very light in weight. Used when optical fiber fault distance is less than 50 m.

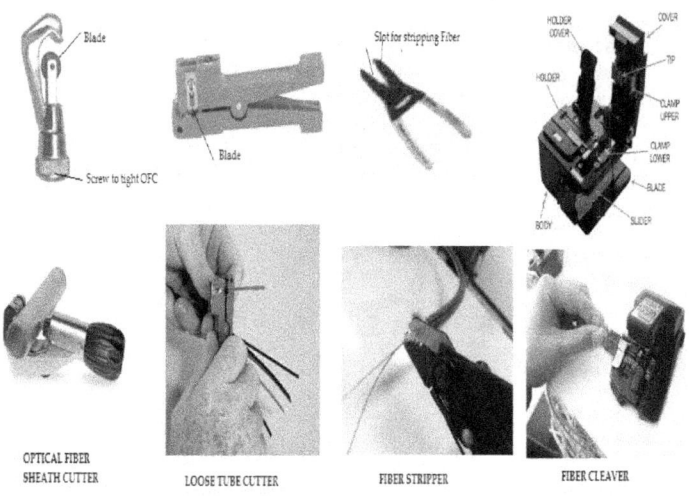

| OPTICAL FIBER SHEATH CUTTER | LOOSE TUBE CUTTER | FIBER STRIPPER | FIBER CLEAVER |

www.ingramcontent.com/pod-product-compliance
Lightning Source LLC
Chambersburg PA
CBHW070130240526
45468CB00002BA/757